Original Korean text by Eun-sook Jo
Illustrations by Yeong-mook Kwon
Korean edition © Dawoolim

This English edition published by big & SMALL in 2015
by arrangement with Dawoolim
English text edited by Joy Cowley
English edition © big & SMALL 2015

ISBN: 978-1-925233 71-1

Printed in Korea

Ah, I Am Full!

Written by Eun-sook Jo
Illustrated by Yeong-mook Kwon
Edited by Joy Cowley

big & SMALL

In the forest, by the lake,
bellflowers swayed in the wind,
soaking in the warm sun's rays.

A food chain starts with a primary energy source, the sun. Plants are called **producers** as they use sunlight, water and nutrients to create their own energy to grow.

A grasshopper nibbled
a bellflower leaf.

Animals are called **consumers** as they get energy to live
from the food they eat. Animals, like the grasshopper,
that only eat plants are called **herbivores**.

8

"Ah, I am full!" said the grasshopper
and away it went with a great hop.

HOP!

The grasshopper landed in a web and a spider wrapped it in silk.

Animals that eat only other animals are called **carnivores**. If an animal eats both plants and other animals, it is called an **omnivore**.

Swing!
Swing!

"Ah, I am full!" said the spider,
swinging from a branch.

A frog snatched the spider
with its long tongue.

"Ah, I am full!" said the frog
and it jumped out of the pond.

JUMP!

A snake slithered out of the grass and grabbed the frog.

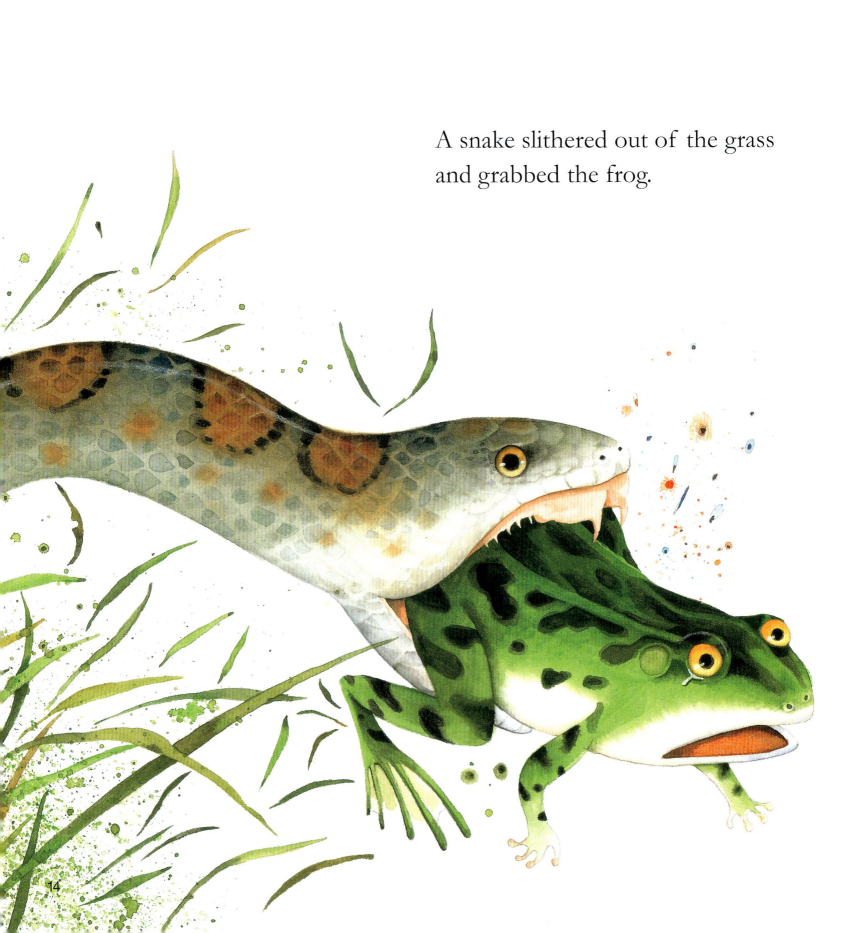

"Ah, I am full!" said the snake, slithering into the bushes.

Slither! Slither!

15

An owl flew down
and pounced on the snake.

"Ah, I am full!" said the owl
and it flew to a tree stump.

Flap!
Flap!

A fox came out of the forest
and snatched the owl.

Sneak! Sneak!

"Ah, I am, full!" said the fox
as it sneaked back
into the forest.

A tiger came out
and jumped on the fox.

20

"Ah! I am full!" roared the tiger.

Roar!
Roar!

An animal that does not have any natural enemies is called a **top predator**. It is at the top of the food chain, but the food chain does not end there.

Wriggle! Wriggle!

The tiger grew old and died.
Flies laid eggs on its body.
The eggs hatched into maggots.

"Ah, we are full!" said the maggots.

Animals that help speed up decay by feeding on decomposing plants and animals are called **detritivores**. Maggots and worms are good examples of detritivores.

Microbes in the soil
broke down the remains.
Soon there was no tiger left.

The soil was full of nutrients.

Microbes, such as bacteria and fungi, eat the microscopic remains of decomposing plants and animals. These microbes are called **decomposers**. They convert these remains into nutrients.

Plants took nutrients from the soil
and the bellflowers swayed in the wind.

Plants grow by using these nutrients together with water and energy from the sun.

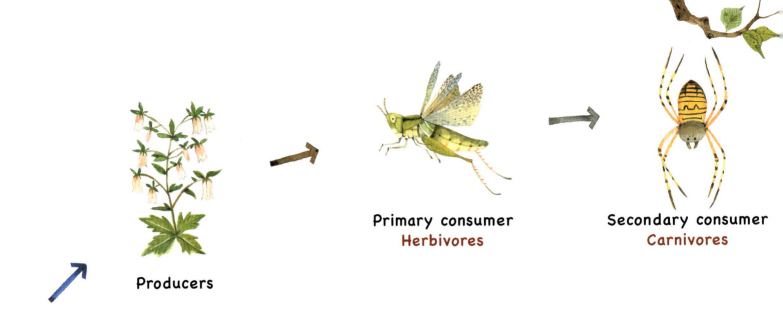

Producers

Primary consumer
Herbivores

Secondary consumer
Carnivores

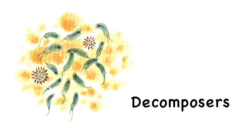

Decomposers

The chain of eating and being eaten has made a full circle.

Top predators

Detritivores

Tertiary consumer

It is amazing!

Quaternary consumer

Ah, I Am Full!

Energy is necessary for living things to grow, and all living things get energy from food. A food chain shows how each living thing gets food, and how nutrients and energy pass between them.

Let's think

Why is it called a **food chain**?

How do plants make their own food?

What are **producers**, **consumers**, **detritivores**, and **decomposers** in the food chain?

What are **carnivores**, **herbivores** and **omnivores**?

What would happen if there were no **microbes**?

Let's Do!

What did you eat for dinner last night? Write down all the individual ingredients that were in your meal. From the ingredients, identify the producers and consumers, as well as which are carnivores, herbivores, omnivores and decomposers. Select one of these ingredients and draw a food chain to show how it ended up in you.
Try it again with a different ingredient.